Astronomy

I can't explain it. Therefore, aliens!

6" x 9" Write or Sketch Journal

Vincent Van Gouache

Published by:
Berhampore Press
Wellington, NZ
Copyright 2017
All Rights Reserved

BerhamporePress@gmail.com

ISBN-13:
978-1545514597

ISBN-10:
1545514593